Mathematical formulae

Second Edition

Compiled by **J.O. Bird and A.J.C. May**

Longman Scientific & Technical
Longman Group UK Limited,
Longman House, Burnt Mill, Harlow,
Essex, CM20 2JE, England
and Associated Companies throughout the world.

First published 1979
Second impression (with additional formulae) 1981
Seventh impression reprinted by Longman Scientific & Technical 1987
Eleventh impression 1991
Second edition 1993

ISBN 0 582 22910 3

British Library Cataloguing in Publication Data
A CIP record for this book is available from the British Library

Set by 4 in 10/11 Compugraphic Times

Printed in Malaysia by VVP

Preface

It is widely recognized that there is little value in requiring students to memorize large amounts of routine material.

This booklet provides students with a readily available reference to the mathematical formulae which they need during their studies.

The mathematical formulae given are relevant to students taking NVQ and gNVQ, BTEC and CGLI courses in Electrical and Electronic Engineering, Mechanical and Production Engineering, Construction, Maritime Studies, General Engineering and Science. The formulae presented are appropriate for the Mathematics for Engineering Modules at National and Higher National levels as well as for students studying GCSE and 'A' level courses.

For this second edition of Mathematical Formulae, material has been added on differential equations, partial differentiation and statistics.

Highbury College, Portsmouth

J.O. Bird
A.J.C. May

Contents

Algebra

Laws of indices:

1. $a^m \times a^n = a^{m+n}$
2. $\dfrac{a^m}{a^n} = a^{m-n}$
3. $(a^m)^n = a^{mn}$
4. $a^{\frac{m}{n}} = \sqrt[n]{a^m}$
5. $a^{-n} = \dfrac{1}{a^n}$
6. $a^0 = 1$

Definition of a logarithm: If $y = a^x$ then $x = \log_a y$

Laws of logarithms:

1. $\log(A \times B) = \log A + \log B$
2. $\log\left(\dfrac{A}{B}\right) = \log A - \log B$
3. $\log A^n = n \times \log A$

Change of base: $\log_a y = \dfrac{\log_b y}{\log_b a}$

$$\ln y = \frac{\lg y}{\lg e} = 2.3026 \lg y$$

Quadratic formula: If $ax^2 + bx + c = 0$

then $x = \dfrac{-b \pm \sqrt{(b^2 - 4ac)}}{2a}$

Partial fractions:

Provided that the numerator $f(x)$ is of less degree than the relevant denominator, the following identities are typical examples of the form of partial fraction used:

$$\frac{f(x)}{(x+a)(x+b)(x+c)} \equiv \frac{A}{(x+a)} + \frac{B}{(x+b)} + \frac{C}{(x+c)}$$

$$\frac{f(x)}{(x-a)^3(x+b)} \equiv \frac{A}{(x-a)} + \frac{B}{(x-a)^2} + \frac{C}{(x-a)^3} + \frac{D}{(x+b)}$$

$$\frac{f(x)}{(ax^2+bx+c)(x-d)} \equiv \frac{Ax+B}{(ax^2+bx+c)} + \frac{C}{(x-d)}$$

Newton-Raphson iterative method:

If r_1 is the approximate value for a real root of the equation $f(x) = 0$, then a closer approximation to the root r_2 is generally given by:

$$r_2 = r_1 - \frac{f(r_1)}{f'(r_1)}$$

Series

Binomial series:

$$(a+b)^n = a^n + na^{n-1}b + \frac{n(n-1)}{2!}a^{n-2}b^2 + \frac{n(n-1)(n-2)}{3!}a^{n-3}b^3 + \ldots$$

$$(1+x)^n = 1 + nx + \frac{n(n-1)}{2!}x^2 + \frac{n(n-1)(n-2)}{3!}x^3 + \ldots$$

(valid for $-1 < x < 1$)

Maclaurin's theorem:

$$f(x) = f(0) + x\,f'(0) + \frac{x^2}{2!}f''(0) + \ldots$$

Taylor's theorem:

$$f(a+h) = f(a) + h\,f'(a) + \frac{h^2}{2!}f''(a) + \ldots$$

Exponential series:

$$e^x = 1 + x + \frac{x^2}{2!} + \frac{x^3}{3!} + \ldots \text{ (valid for all values of } x\text{)}$$

Logarithmic series:

$$\ln(1+x) = x - \frac{x^2}{2} + \frac{x^3}{3} - \frac{x^4}{4} + \ldots \text{ (valid if } -1 < x \leqslant 1\text{)}$$

Trigonometrical series:

$$\sin x = x - \frac{x^3}{3!} + \frac{x^5}{5!} - \frac{x^7}{7!} + \ldots \text{ (valid for all values of } x\text{)}$$

$$\cos x = 1 - \frac{x^2}{2!} + \frac{x^4}{4!} - \frac{x^6}{6!} + \ldots \text{ (valid for all values of } x\text{)}$$

Hyperbolic series:

$$\sinh x = x + \frac{x^3}{3!} + \frac{x^5}{5!} + \frac{x^7}{7!} + \ldots \text{ (valid for all values of } x\text{)}$$

$$\cosh x = 1 + \frac{x^2}{2!} + \frac{x^4}{4!} + \frac{x^6}{6!} + \ldots \text{ (valid for all values of } x\text{)}$$

Arithmetic progression:

If a = first term, d = common difference and n = number of terms then the arithmetic progression is: $a, a + d, a + 2d, \ldots$

The n'th term is: $a + (n - 1)d$

Sum of n terms, $S_n = \frac{n}{2}[2a + (n - 1)d] = \frac{n}{2}(a + l)$ where l is the last term.

Geometric progression:

If a = first term, r = common ratio and n = number of terms then the geometric progression is: $a, ar, ar^2, \ldots$

The n'th term is ar^{n-1}

$$\text{Sum of } n \text{ terms, } S_n = \frac{a(1 - r^n)}{(1 - r)} \text{ or } \frac{a(r^n - 1)}{(r - 1)}$$

$$\text{If } -1 < r < 1,\ S_\infty = \frac{a}{(1 - r)}$$

Complex numbers

$z = a + jb = r(\cos\theta + j\sin\theta) = r\angle\theta = r\,e^{j\theta}$, where $j^2 = -1$

Modulus, $r = |z| = \sqrt{(a^2 + b^2)}$

Argument, $\theta = \arg z = \arctan\frac{b}{a}$

Addition: $(a + jb) + (c + jd) = (a + c) + j(b + d)$

Subtraction: $(a + jb) - (c + jd) = (a - c) + j(b - d)$

Complex equations: If $m + jn = p + jq$ then $m = p$ and $n = q$

Multiplication: $z_1 z_2 = r_1 r_2 \angle(\theta_1 + \theta_2)$

Division: $\frac{z_1}{z_2} = \frac{r_1}{r_2}\angle(\theta_1 - \theta_2)$

De Moivre's theorem: $[r\angle\theta]^n = r^n\angle n\theta = r^n(\cos n\theta + j\sin n\theta)$

cos $n\theta$ and sin $n\theta$ in terms of powers of cos θ and sin θ:

$$\cos n\theta = \cos^n\theta - \frac{n(n-1)}{2!}\cos^{n-2}\theta\sin^2\theta + \frac{n(n-1)(n-2)(n-3)}{4!}\cos^{n-4}\theta\sin^4\theta - \ldots$$

$$\sin n\theta = n\cos^{n-1}\theta\sin\theta - \frac{n(n-1)(n-2)}{3!}\cos^{n-3}\theta\sin^3\theta + \frac{n(n-1)(n-2)(n-3)(n-4)}{5!}\cos^{n-5}\theta\sin^5\theta - \ldots$$

$\cos^n\theta$ and $\sin^n\theta$ in terms of sines and cosines of multiples of θ:

If $z = (\cos\theta + j\sin\theta)$ then:

$$\cos^n\theta = \frac{1}{2^n}\left(z + \frac{1}{z}\right)^n \qquad \left(z^n + \frac{1}{z^n}\right) = 2\cos n\theta$$

$$\sin^n\theta = \frac{1}{j^n 2^n}\left(z - \frac{1}{z}\right)^n \qquad \left(z^n - \frac{1}{z^n}\right) = 2j\sin n\theta$$

Geometry

Equations of a straight line:

(1) $y = mx + c$

(2) $y - y_1 = m(x - x_1)$

Reduction of equations to linear form:

	Vertical axis	Gradient	Horizontal axis	Intercept on vertical axis
If $y = ax^n$ then	$\lg y$	$= n$	$\lg x$	$+ \lg a$
$y = ab^x$ then	$\lg y$	$= \lg b$	x	$+ \lg a$
$y = ae^{kx}$ then	$\ln y$	$= k$	x	$+ \ln a$
$y = ax^n + bx^{n-1}$ then	$\dfrac{y}{x^{n-1}}$	$= a$	x	$+ b$

Equation of a circle, centre at origin, radius r:

$x^2 + y^2 = r^2$

Equations of a parabola:

(1) $y = ax^2 + bx + c$

(2) $y^2 = 4ac$

Equation of an ellipse, centre at origin, semi-axes a and b:

$$\frac{x^2}{a^2} + \frac{y^2}{b^2} = 1$$

Equation of a hyperbola:

$$\frac{x^2}{a^2} - \frac{y^2}{b^2} = 1$$

Equation of a rectangular hyperbola:

$$xy = c^2$$

Theorem of Pythagoras:

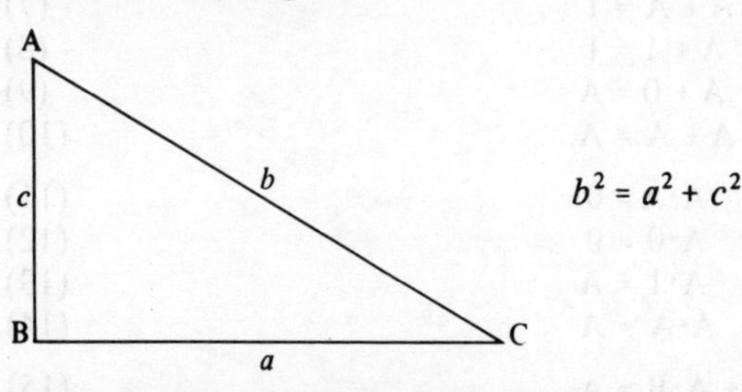

$$b^2 = a^2 + c^2$$

Centroids of common shapes

The centroid of:
(a) a *rectangle* lies on the intersection of the diagonals;
(b) a *triangle*, of perpendicular height h, lies at a point $\frac{h}{3}$ from the base;
(c) a *circle* lies at its centre;
(d) a *semicircle*, of radius r, lies on the centre line at a distance $\frac{4r}{3\pi}$ from the diameter.

First moment of area:

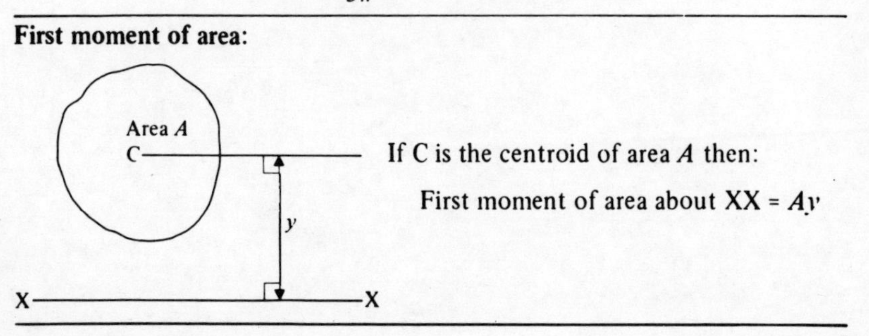

If C is the centroid of area A then:

First moment of area about XX = Ay

Radian measure:

2π radians = 360 degrees.

Boolean algebra

Laws and rules of Boolean algebra

Commutative Laws:

$$A + B = B + A \quad (1)$$
$$A \cdot B = B \cdot A \quad (2)$$

Associative Laws:

$$A + B + C = (A + B) + C \quad (3)$$
$$A \cdot B \cdot C = (A \cdot B) \cdot C \quad (4)$$

Distributive Laws:

$$A \cdot (B + C) = A \cdot B + A \cdot C \quad (5)$$
$$A + (B \cdot C) = (A + B) \cdot (A + C) \quad (6)$$

Sum rules:

$$A + \bar{A} = 1 \quad (7)$$
$$A + 1 = 1 \quad (8)$$
$$A + 0 = A \quad (9)$$
$$A + A = A \quad (10)$$

Product rules:

$$A \cdot \bar{A} = 0 \quad (11)$$
$$A \cdot 0 = 0 \quad (12)$$
$$A \cdot 1 = A \quad (13)$$
$$A \cdot A = A \quad (14)$$

Absorption rules:

$$A + A \cdot B = A \quad (15)$$
$$A \cdot (A + B) = A \quad (16)$$
$$A + \bar{A} \cdot B = A + B \quad (17)$$

De Morgan's Laws:

$$\overline{A + B} = \bar{A} \cdot \bar{B} \quad (18)$$
$$\overline{A \cdot B} = \bar{A} + \bar{B} \quad (19)$$

Mensuration

Areas of plane figures

1. Rectangle

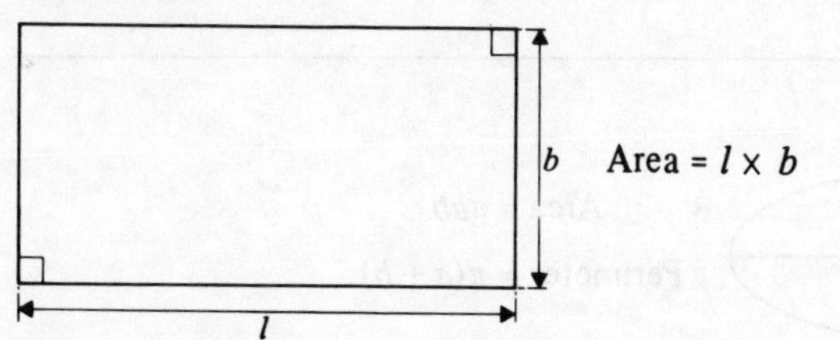

$$\text{Area} = l \times b$$

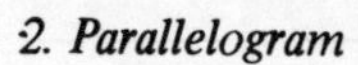

2. Parallelogram

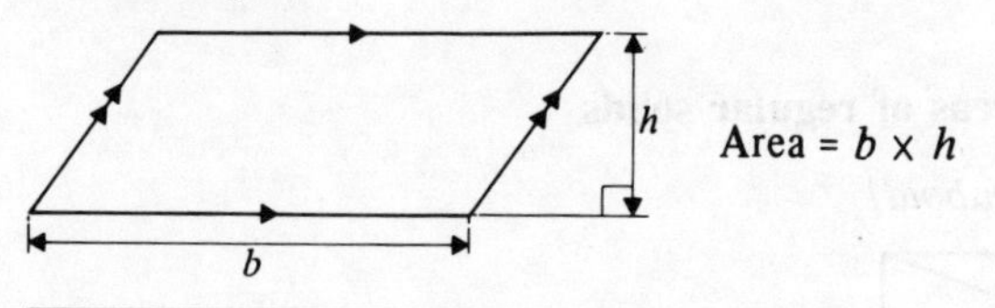

$$\text{Area} = b \times h$$

3. Trapezium

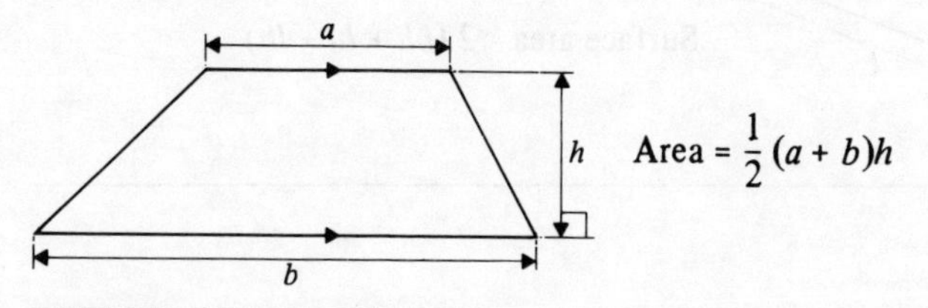

$$\text{Area} = \frac{1}{2}(a + b)h$$

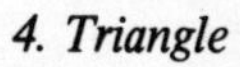

4. Triangle

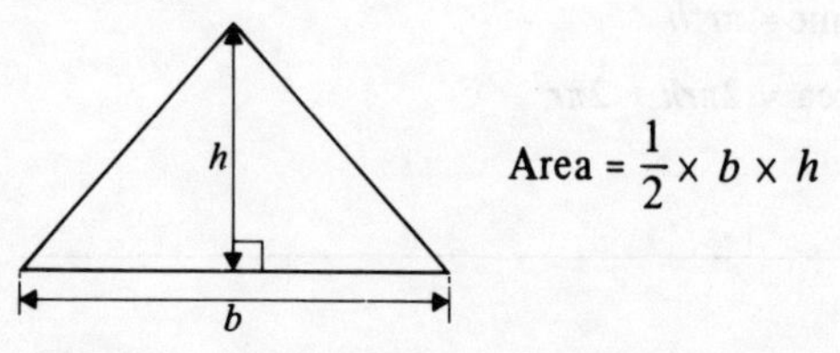

$$\text{Area} = \frac{1}{2} \times b \times h$$

5. Circle

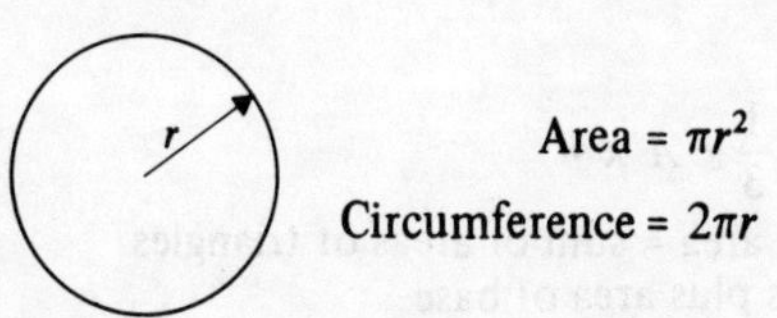

$$\text{Area} = \pi r^2$$

$$\text{Circumference} = 2\pi r$$

6. Sector of a circle

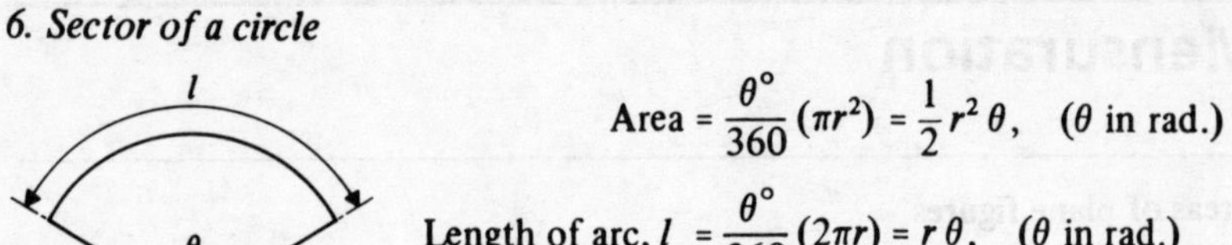

$$\text{Area} = \frac{\theta°}{360}(\pi r^2) = \frac{1}{2}r^2\theta, \quad (\theta \text{ in rad.})$$

$$\text{Length of arc, } l = \frac{\theta°}{360}(2\pi r) = r\theta, \quad (\theta \text{ in rad.})$$

7. Ellipse

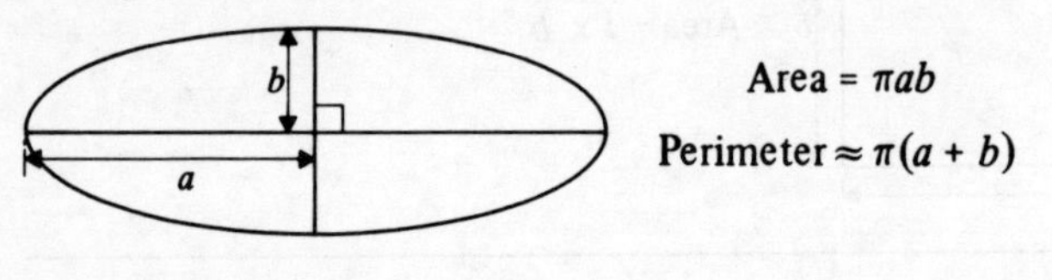

$$\text{Area} = \pi ab$$

$$\text{Perimeter} \approx \pi(a + b)$$

Volumes and surface areas of regular solids

1. Rectangular prism (or cuboid)

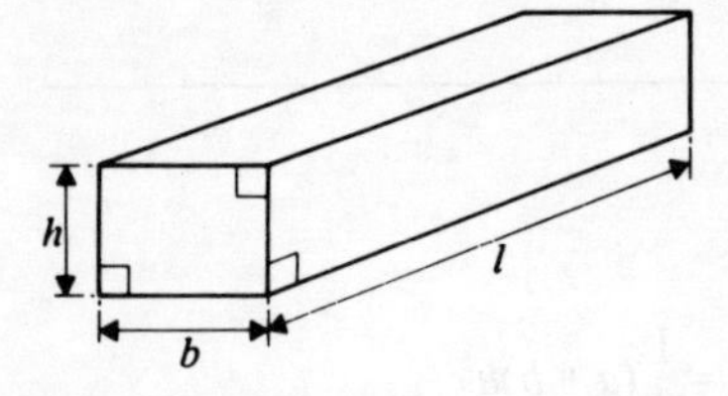

$$\text{Volume} = l \times b \times h$$

$$\text{Surface area} = 2\,(bh + hl + lb)$$

2. Cylinder

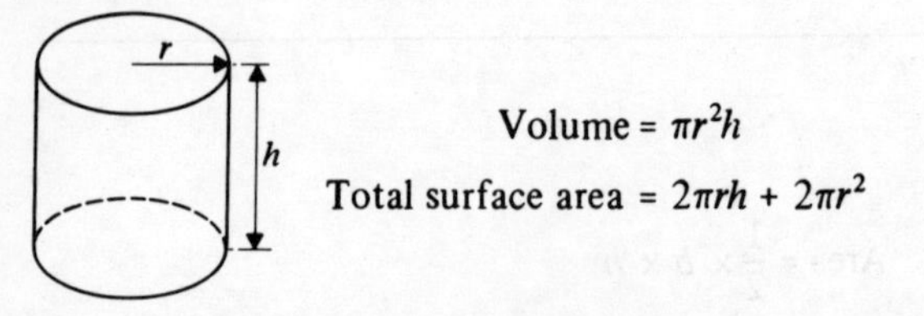

$$\text{Volume} = \pi r^2 h$$

$$\text{Total surface area} = 2\pi rh + 2\pi r^2$$

3. Pyramid

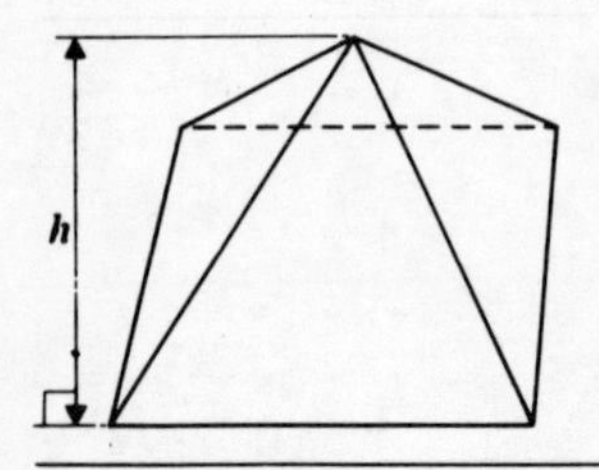

If area of base = A and perpendicular height = h then:

$$\text{Volume} = \frac{1}{3} \times A \times h$$

Total surface area = sum of areas of triangles forming sides plus area of base.

4. Cone

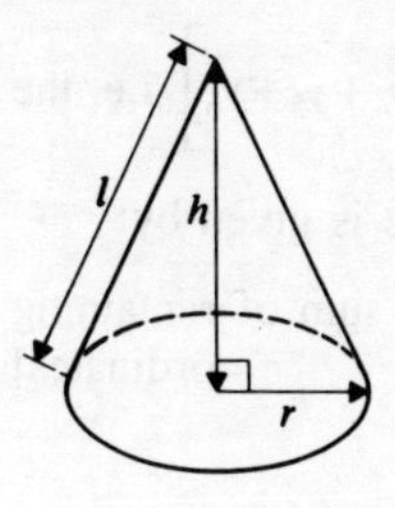

$$\text{Volume} = \frac{1}{3}\pi r^2 h$$

$$\text{Curved surface area} = \pi r l$$

$$\text{Total surface area} = \pi r l + \pi r^2$$

---

5. Frustum of a cone

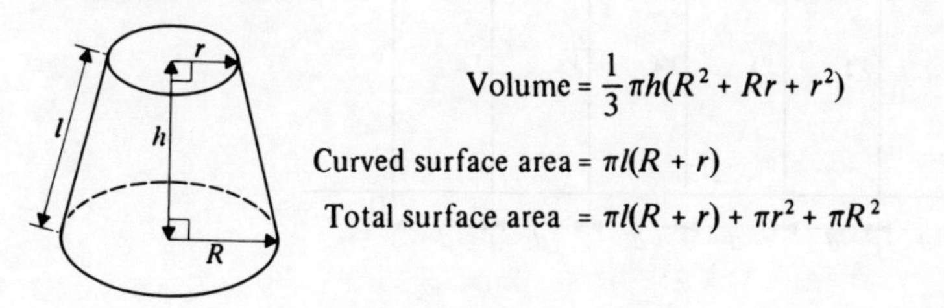

$$\text{Volume} = \frac{1}{3}\pi h(R^2 + Rr + r^2)$$

$$\text{Curved surface area} = \pi l(R + r)$$

$$\text{Total surface area} = \pi l(R + r) + \pi r^2 + \pi R^2$$

---

6. Sphere

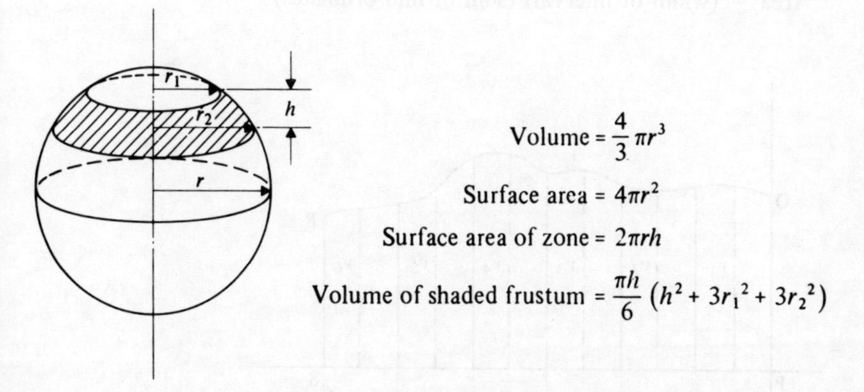

$$\text{Volume} = \frac{4}{3}\pi r^3$$

$$\text{Surface area} = 4\pi r^2$$

$$\text{Surface area of zone} = 2\pi r h$$

$$\text{Volume of shaded frustum} = \frac{\pi h}{6}\left(h^2 + 3{r_1}^2 + 3{r_2}^2\right)$$

Areas of irregular figures by approximate methods

Trapezoidal rule

From Fig. 1, area of ABCD $= d\left[\left(\frac{y_1+y_7}{2}\right)+y_2+y_3+y_4+y_5+y_6\right]$ i.e. the trapezoidal rule states that the area of an irregular figure is given by:

Area = (width of interval) [$\frac{1}{2}$ (first + last ordinates) + sum of remaining ordinates]

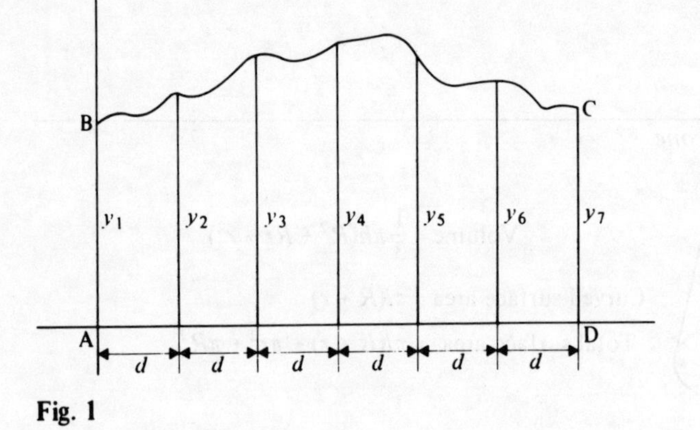

Fig. 1

Mid-ordinate rule

From Fig. 2, area of PQRS $= d[y_1+y_2+y_3+y_4+y_5+y_6]$, i.e. the mid-ordinate rule states that the area of an irregular figure is given by:

Area = (width of interval) (sum of mid-ordinates)

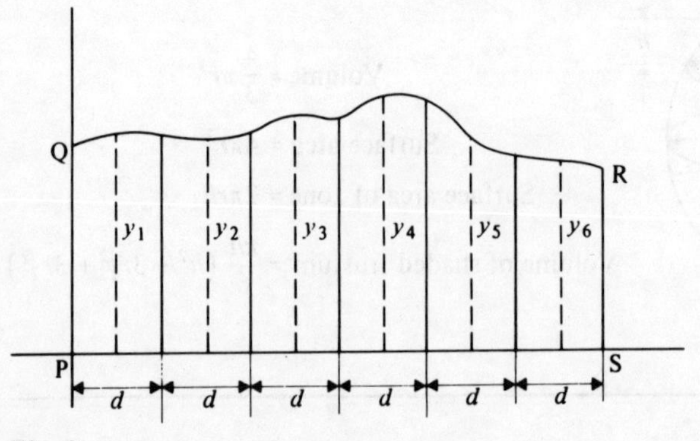

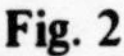

Fig. 2

Simpson's rule

To find an area such as ABCD of Fig. 1 the base AD must be divided into an even number of strips of equal width d, thus producing an odd number of ordinates, in this case 7.

$$\text{Area of ABCD} = \frac{d}{3}\,[(y_1+y_7)+4(y_2+y_4+y_6)+2(y_3+y_5)]$$

i.e. Simpson's rule states that the area of an irregular figure is given by:

$$\text{Area} = \tfrac{1}{3}(\text{width of interval})\left[\begin{pmatrix}\text{first}+\text{last}\\ \text{ordinate}\end{pmatrix} + 4\begin{pmatrix}\text{sum of even}\\ \text{ordinates}\end{pmatrix} + 2\begin{pmatrix}\text{sum of remaining}\\ \text{odd ordinates}\end{pmatrix}\right]$$

Mean or average value of a waveform

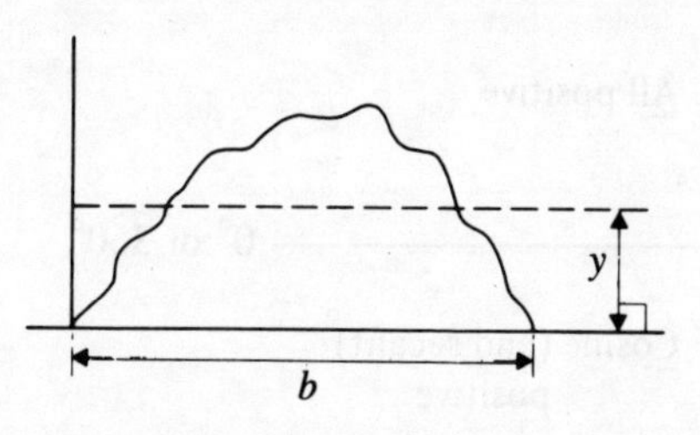

Mean or average value, y

$$= \frac{\text{area under curve}}{\text{length of base } (b)}$$

$$= \frac{\text{sum of mid-ordinates}}{\text{number of mid-ordinates}}$$

Prismoidal rule for finding volumes

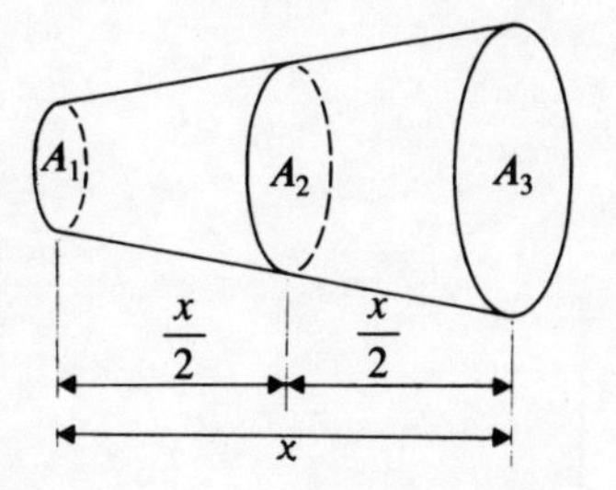

$$\text{Volume} = \frac{x}{6}\,[A_1+4A_2+A_3]$$

Trigonometry

Identities: $\sec\theta = \dfrac{1}{\cos\theta}$, $\operatorname{cosec}\theta = \dfrac{1}{\sin\theta}$, $\cot\theta = \dfrac{1}{\tan\theta}$, $\tan\theta = \dfrac{\sin\theta}{\cos\theta}$

$$\cos^2\theta + \sin^2\theta = 1$$
$$1 + \tan^2\theta = \sec^2\theta$$
$$\cot^2\theta + 1 = \operatorname{cosec}^2\theta$$
$$\sin(-\theta) = -\sin\theta$$
$$\cos(-\theta) = +\cos\theta$$
$$\tan(-\theta) = -\tan\theta$$

Trigonometric ratios for angles of any magnitude

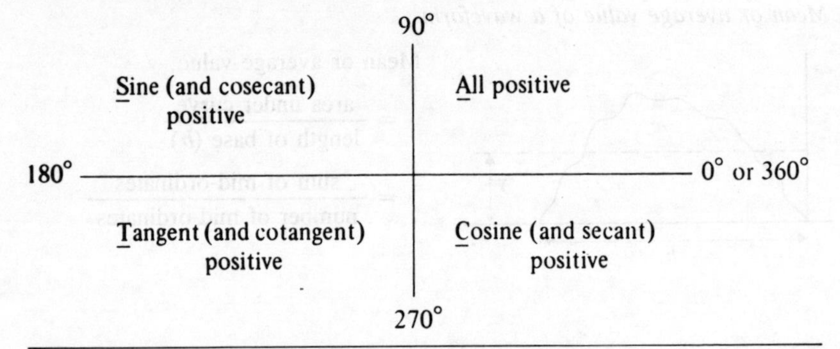

Compound angle addition and subtraction formulae

$$\sin(A+B) = \sin A\cos B + \cos A\sin B$$
$$\sin(A-B) = \sin A\cos B - \cos A\sin B$$
$$\cos(A+B) = \cos A\cos B - \sin A\sin B$$
$$\cos(A-B) = \cos A\cos B + \sin A\sin B$$
$$\tan(A+B) = \frac{\tan A + \tan B}{1 - \tan A\tan B}$$
$$\tan(A-B) = \frac{\tan A - \tan B}{1 + \tan A\tan B}$$

If $R\sin(\omega t+\alpha) = a\sin\omega t + b\cos\omega t$,

then $a = R\cos\alpha$, $b = R\sin\alpha$, $R = \sqrt{(a^2+b^2)}$ and $\alpha = \arctan\dfrac{b}{a}$

Double angles

$$\sin 2A = 2 \sin A \cos A$$

$$\cos 2A = \cos^2 A - \sin^2 A = 2 \cos^2 A - 1 = 1 - 2 \sin^2 A$$

$$\tan 2A = \frac{2 \tan A}{1 - \tan^2 A}$$

Half angles

If $\tan \frac{x}{2} = t$ then $\sin x = \dfrac{2t}{1 + t^2}$

$$\cos x = \frac{1 - t^2}{1 + t^2}$$

$$\tan x = \frac{2t}{1 - t^2}$$

Products of sines and cosines into sums or differences

$$\sin A \cos B = \tfrac{1}{2} [\sin (A + B) + \sin (A - B)]$$

$$\cos A \sin B = \tfrac{1}{2} [\sin (A + B) - \sin (A - B)]$$

$$\cos A \cos B = \tfrac{1}{2} [\cos (A + B) + \cos (A - B)]$$

$$\sin A \sin B = -\tfrac{1}{2} [\cos (A + B) - \cos (A - B)]$$

Sums or differences of sines and cosines into products

$$\sin x + \sin y = 2 \sin\left(\frac{x + y}{2}\right) \cos\left(\frac{x - y}{2}\right)$$

$$\sin x - \sin y = 2 \cos\left(\frac{x + y}{2}\right) \sin\left(\frac{x - y}{2}\right)$$

$$\cos x + \cos y = 2 \cos\left(\frac{x + y}{2}\right) \cos\left(\frac{x - y}{2}\right)$$

$$\cos x - \cos y = -2 \sin\left(\frac{x + y}{2}\right) \sin\left(\frac{x - y}{2}\right)$$

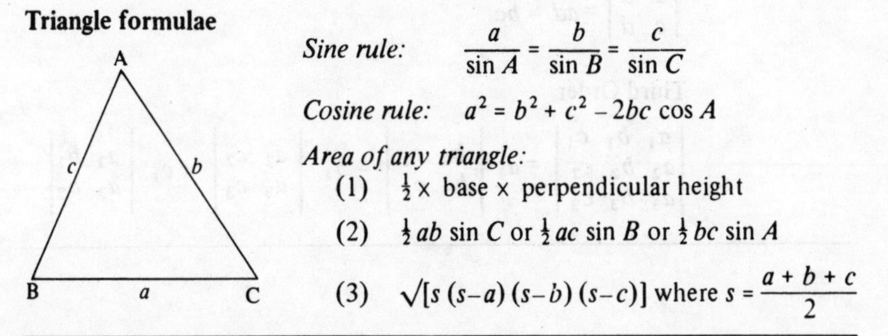

Triangle formulae

Sine rule: $\dfrac{a}{\sin A} = \dfrac{b}{\sin B} = \dfrac{c}{\sin C}$

Cosine rule: $a^2 = b^2 + c^2 - 2bc \cos A$

Area of any triangle:

(1) $\tfrac{1}{2} \times$ base $\times$ perpendicular height

(2) $\tfrac{1}{2} ab \sin C$ or $\tfrac{1}{2} ac \sin B$ or $\tfrac{1}{2} bc \sin A$

(3) $\sqrt{[s (s-a) (s-b) (s-c)]}$ where $s = \dfrac{a + b + c}{2}$

Hyperbolic functions

Definitions:

$$\sinh x = \frac{e^x - e^{-x}}{2} \qquad \text{cosech}\, x = \frac{1}{\sinh x} = \frac{2}{e^x - e^{-x}}$$

$$\cosh x = \frac{e^x + e^{-x}}{2} \qquad \text{sech}\, x = \frac{1}{\cosh x} = \frac{2}{e^x + e^{-x}}$$

$$\tanh x = \frac{e^x - e^{-x}}{e^x + e^{-x}} \qquad \coth x = \frac{1}{\tanh x} = \frac{e^x + e^{-x}}{e^x - e^{-x}}$$

Identities:

$$\cosh^2 x - \sinh^2 x = 1$$

$$1 - \tanh^2 x = \text{sech}^2 x$$

$$\coth^2 x - 1 = \text{cosech}^2 x$$

Matrices and determinants

Matrices:

If $A = \begin{pmatrix} a & b \\ c & d \end{pmatrix}$ and $B = \begin{pmatrix} e & f \\ g & h \end{pmatrix}$

then $A + B = \begin{pmatrix} a+e & b+f \\ c+g & d+h \end{pmatrix}$

$$A - B = \begin{pmatrix} a-e & b-f \\ c-g & d-h \end{pmatrix}$$

$$A \times B = \begin{pmatrix} ae+bg & af+bh \\ ce+dg & cf+dh \end{pmatrix}$$

$$A^{-1} = \frac{1}{ad-bc}\begin{pmatrix} d & -b \\ -c & a \end{pmatrix}$$

Determinants: Second Order

$$\begin{vmatrix} a & b \\ c & d \end{vmatrix} = ad - bc$$

Third Order

$$\begin{vmatrix} a_1 & b_1 & c_1 \\ a_2 & b_2 & c_2 \\ a_3 & b_3 & c_3 \end{vmatrix} = a_1 \begin{vmatrix} b_2 & c_2 \\ b_3 & c_3 \end{vmatrix} - b_1 \begin{vmatrix} a_2 & c_2 \\ a_3 & c_3 \end{vmatrix} + c_1 \begin{vmatrix} a_2 & b_2 \\ a_3 & b_3 \end{vmatrix}$$

Differential calculus

Standard derivatives:

y or $f(x)$	$\frac{dy}{dx}$ or $f'(x)$
(1) ax^n	$an\,x^{n-1}$
(2) $\sin ax$	$a \cos ax$
(3) $\cos ax$	$-a \sin ax$
(4) $\tan ax$	$a \sec^2 ax$
(5) $\sec ax$	$a \sec ax \tan ax$
(6) $\operatorname{cosec} ax$	$-a \operatorname{cosec} ax \cot ax$
(7) $\cot ax$	$-a \operatorname{cosec}^2 ax$
(8) e^{ax}	ae^{ax}
(9) $\ln ax$	$\frac{1}{x}$
(10) a^x	$a^x \ln a$
(11) $\arcsin\frac{x}{a}$	$\frac{1}{\sqrt{(a^2 - x^2)}}$
(12) $\arccos\frac{x}{a}$	$-\frac{1}{\sqrt{(a^2 - x^2)}}$
(13) $\arctan\frac{x}{a}$	$\frac{a}{a^2 + x^2}$
(14) $\operatorname{arcsec}\frac{x}{a}$	$\frac{a}{x\sqrt{(x^2 - a^2)}}$
(15) $\operatorname{arccosec}\frac{x}{a}$	$-\frac{a}{x\sqrt{(x^2 - a^2)}}$
(16) $\operatorname{arccot}\frac{x}{a}$	$-\frac{a}{a^2 + x^2}$
(17) $\sinh ax$	$a \cosh ax$
(18) $\cosh ax$	$a \sinh ax$
(19) $\tanh ax$	$a \operatorname{sech}^2 ax$
(20) $\operatorname{sech} ax$	$-a \operatorname{sech} ax \tanh ax$
(21) $\operatorname{cosech} ax$	$-a \operatorname{cosech} ax \coth ax$
(22) $\coth ax$	$-a \operatorname{cosech}^2 ax$

y or $f(x)$	$\frac{dy}{dx}$ or $f'(x)$
(23) $\text{arsinh}\,\frac{x}{a}$	$\frac{1}{\sqrt{(x^2+a^2)}}$
(24) $\text{arcosh}\,\frac{x}{a}$	$\frac{1}{\sqrt{(x^2-a^2)}}$
(25) $\text{artanh}\,\frac{x}{a}$	$\frac{a}{a^2-x^2}$
(26) $\text{arsech}\,\frac{x}{a}$	$\frac{a}{-x\sqrt{(a^2-x^2)}}$
(27) $\text{arcosech}\,\frac{x}{a}$	$-\frac{a}{x\sqrt{(x^2+a^2)}}$
(28) $\text{arcoth}\,\frac{x}{a}$	$\frac{a}{a^2-x^2}$

Product rule: **When $y = uv$ and u and v are functions of x then:**

$$\frac{dy}{dx} = v\frac{du}{dx} + u\frac{dv}{dx}$$

Quotient rule: When $y = \frac{u}{v}$ and u and v are functions of x then:

$$\frac{dy}{dx} = \frac{v\frac{du}{dx} - u\frac{dv}{dx}}{v^2}$$

Maximum and minimum values:

If $y = f(x)$ then $\frac{dy}{dx} = 0$ for stationary points.

Let a solution of $\frac{dy}{dx} = 0$ be $x = a$.

If the value of $\frac{d^2y}{dx^2}$ when $x = a$ is:

positive, the point is a *minimum* value,
negative, the point is a *maximum* value,
zero, the point is a *point of inflexion.*

Function of a function rule (i.e. chain rule):

If u is a function of x then: $\frac{dy}{dx} = \frac{dy}{du} \times \frac{du}{dx}$

Implicit differentiation:

$$\frac{d}{dx}[f(y)] = \frac{d}{dy}[f(y)] \times \frac{dy}{dx}$$

Partial differentiation:

Total differential

If $z = f(u, v, \ldots)$, then the total differential,

dz, is given by:

$$dz = \frac{\partial z}{\partial u} du + \frac{\partial z}{\partial v} dv + \ldots$$

Rate of change

If $z = f(u, v, \ldots)$ and $\frac{du}{dt}, \frac{dv}{dt}, \ldots$ denote

the rate of change of u, v, ... respectively,

then the rate of change of z, $\frac{dz}{dt}$, is given by:

$$\frac{dz}{dt} = \frac{\partial z}{\partial u}.\frac{du}{dt} + \frac{\partial z}{\partial v}.\frac{dv}{dt} + \ldots$$

Small changes

If $z = f(x, y, \ldots)$ and $\delta x, \delta y, \ldots$ denote small changes in x, y, ... respectively, then the corresponding change δz in z is given by:

$$\delta z \approx \frac{\partial z}{\partial x} \delta x + \frac{\partial z}{\partial y} \delta y + \ldots$$

Integral calculus

Standard integrals

y	$\int y\,dx$	
(1) ax^n	$a\dfrac{x^{n+1}}{n+1} + c$ (except where $n = -1$)	
(2) $\cos ax$	$\dfrac{1}{a}\sin ax + c$	
(3) $\sin ax$	$-\dfrac{1}{a}\cos ax + c$	
(4) $\sec^2 ax$	$\dfrac{1}{a}\tan ax + c$	
(5) $\operatorname{cosec}^2 ax$	$-\dfrac{1}{a}\cot ax + c$	
(6) $\operatorname{cosec} ax \cot ax$	$-\dfrac{1}{a}\operatorname{cosec} ax + c$	
(7) $\sec ax \tan ax$	$\dfrac{1}{a}\sec ax + c$	
(8) e^{ax}	$\dfrac{1}{a}e^{ax} + c$	
(9) $\dfrac{1}{x}$	$\ln x + c$	
(10) $\tan ax$	$\dfrac{1}{a}\ln(\sec ax) + c$	
(11) $\cos^2 x$	$\dfrac{1}{2}\left(x + \dfrac{\sin 2x}{2}\right) + c$	(use $\cos 2x = 2\cos^2 x - 1$)
(12) $\sin^2 x$	$\dfrac{1}{2}\left(x - \dfrac{\sin 2x}{2}\right) + c$	(use $\cos 2x = 1 - 2\sin^2 x$)
(13) $\tan^2 x$	$\tan x - x + c$	(use $1 + \tan^2 x = \sec^2 x$)
(14) $\cot^2 x$	$-\cot x - x + c$	(use $\cot^2 x + 1 = \operatorname{cosec}^2 x$)
(15) $\dfrac{1}{x^2 - a^2}$	$\dfrac{1}{2a}\ln\left(\dfrac{x-a}{x+a}\right) + c$	(use partial fractions)

y	$\int y\,dx$
(16) $\dfrac{1}{a^2 - x^2}$	$\dfrac{1}{2a}\ln\left(\dfrac{a+x}{a-x}\right) + c$ (use partial fractions) or $\dfrac{1}{a}\operatorname{artanh}\dfrac{x}{a} + c$ (use $x = a\tanh\theta$ substitution)
(17) $\dfrac{1}{\sqrt{(a^2 - x^2)}}$	$\arcsin\dfrac{x}{a} + c$ (use $x = a\sin\theta$ substitution)
(18) $\sqrt{(a^2 - x^2)}$	$\dfrac{a^2}{2}\arcsin\dfrac{x}{a} + \dfrac{x}{2}\sqrt{(a^2 - x^2)} + c$ (use $x = a\sin\theta$ substitution)
(19) $\dfrac{1}{a^2 + x^2}$	$\dfrac{1}{a}\arctan\dfrac{x}{a} + c$ (use $x = a\tan\theta$ substitution)
(20) $\dfrac{1}{\sqrt{(x^2 + a^2)}}$	$\operatorname{arsinh}\dfrac{x}{a} + c$ (use $x = a\sinh\theta$ substitution) or $\ln\left[\dfrac{x + \sqrt{(x^2 + a^2)}}{a}\right] + c$
(21) $\sqrt{(x^2 + a^2)}$	$\dfrac{a^2}{2}\operatorname{arsinh}\dfrac{x}{a} + \dfrac{x}{2}\sqrt{(x^2 + a^2)} + c$ (use $x = a\sinh\theta$ substitution)
(22) $\dfrac{1}{\sqrt{(x^2 - a^2)}}$	$\operatorname{arcosh}\dfrac{x}{a} + c$ (use $x = a\cosh\theta$ substitution) or $\ln\left[\dfrac{x + \sqrt{(x^2 - a^2)}}{a}\right] + c$
(23) $\sqrt{(x^2 - a^2)}$	$\dfrac{x}{2}\sqrt{(x^2 - a^2)} - \dfrac{a^2}{2}\operatorname{arcosh}\dfrac{x}{a} + c$ (use $x = a\cosh\theta$ substitution)

Integration by parts

If u and v are both functions of x then:

$$\int u\frac{dv}{dx}\,dx = uv - \int v\frac{du}{dx}\,dx.$$

Applications of integration

(1) Area under curve

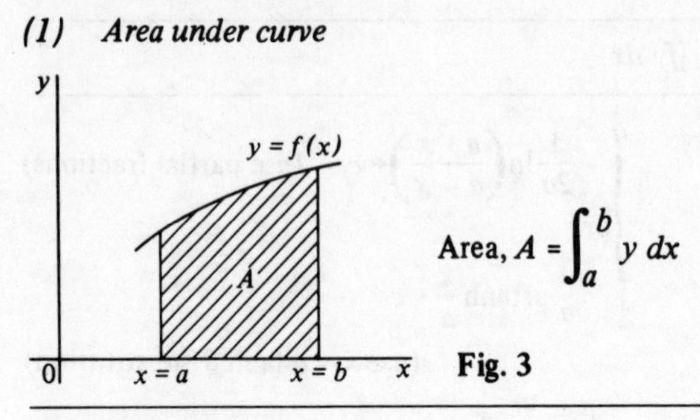

$$\text{Area, } A = \int_a^b y\,dx$$

Fig. 3

---

(2) Mean value

With reference to Fig. 3, the mean value of $y = f(x)$ between $x = a$ and $x = b$ is given by:

$$\text{mean value} = \frac{1}{b-a}\int_a^b y\,dx$$

---

(3) Volume of revolution

With reference to Fig. 3, the volume of revolution, V, obtained by rotating area A through one revolution is given by:

(i) $V = \int_a^b \pi y^2 dx$, about the x-axis, and

(ii) $V = \int_a^b 2\pi\, xy\, dx$, about the y-axis

---

(4) Root mean square value

With reference to Fig. 3, the r.m.s. value of $y = f(x)$ over the range $x = a$ to $x = b$ is given by:

$$\text{r.m.s. value} = \sqrt{\left[\frac{1}{b-a}\int_a^b y^2\,dx\right]}$$

---

(5) Centroids

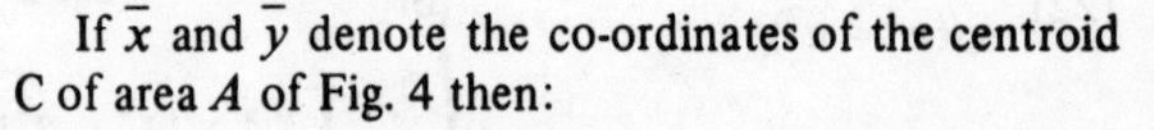
If $\bar{x}$ and $\bar{y}$ denote the co-ordinates of the centroid C of area A of Fig. 4 then:

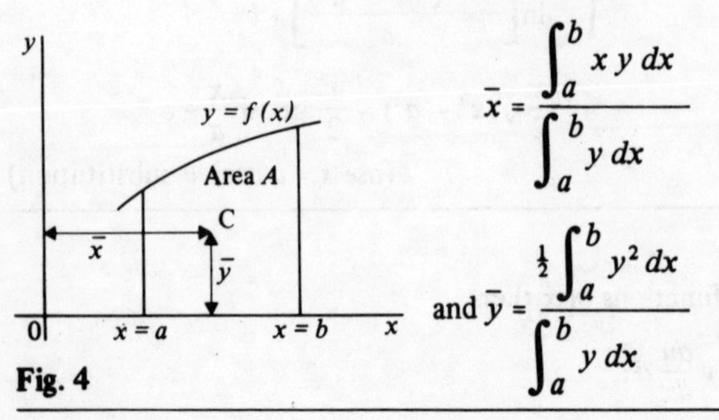

Fig. 4

$$\bar{x} = \frac{\int_a^b x\,y\,dx}{\int_a^b y\,dx}$$

$$\text{and } \bar{y} = \frac{\frac{1}{2}\int_a^b y^2\,dx}{\int_a^b y\,dx}$$

Theorem of Pappus (or Guldinus)

With reference to Fig. 4, when the curve $y = f(x)$ is rotated one revolution about the x-axis between the limits $x = a$ and $x = b$, the volume V generated is given by:

Volume, $V = 2\pi A\bar{y}$

(6) *Second moment of area*

Shape	Position of axis	Second moment of area	Radius of gyration k
Rectangle length l breadth b area A	(1) Coinciding with b	$\frac{bl^3}{3}$ or $A\frac{l^2}{3}$	$\frac{l}{\sqrt{3}}$
	(2) Coinciding with l	$\frac{lb^3}{3}$ or $A\frac{b^2}{3}$	$\frac{b}{\sqrt{3}}$
	(3) Through centroid, parallel to b	$\frac{bl^3}{12}$ or $A\frac{l^2}{12}$	$\frac{l}{\sqrt{12}}$ or $\frac{l}{2\sqrt{3}}$
Triangle perpendicular height h base b area A	(1) Coinciding with base	$\frac{bh^3}{12}$ or $A\frac{h^2}{6}$	$\frac{h}{\sqrt{6}}$
	(2) Through centroid, parallel to base	$\frac{bh^3}{36}$ or $A\frac{h^2}{18}$	$\frac{h}{\sqrt{18}}$ or $\frac{h}{3\sqrt{2}}$
	(3) Through vertex, parallel to base	$\frac{bh^3}{4}$ or $A\frac{h^2}{2}$	$\frac{h}{\sqrt{2}}$
Circle radius r area A	(1) Through centre, perpendicular to plane (i.e. polar axis).	$\frac{\pi r^4}{2}$ or $A\frac{r^2}{2}$	$\frac{r}{\sqrt{2}}$
	(2) Coinciding with diameter	$\frac{\pi r^4}{4}$ or $A\frac{r^2}{4}$	$\frac{r}{2}$
	(3) About a tangent	$\frac{5\pi}{4}r^4$ or $\frac{5}{4}Ar^2$	$\frac{\sqrt{5}}{2}r$
Semicircle radius r area A	Coinciding with diameter.	$\frac{\pi r^4}{8}$ or $A\frac{r^2}{4}$	$\frac{r}{2}$

Parallel axis theorem

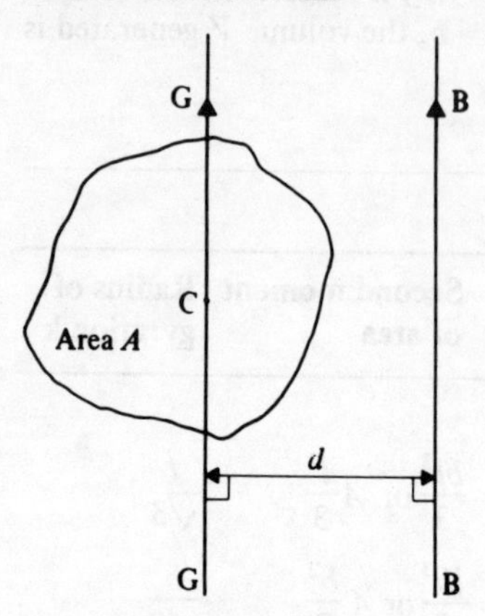

If C is the centroid of area A then:

$$A\,k_{BB}^2 = A\,k_{GG}^2 + A\,d^2$$

or $$k_{BB}^2 = k_{GG}^2 + d^2$$

Perpendicular axis theorem

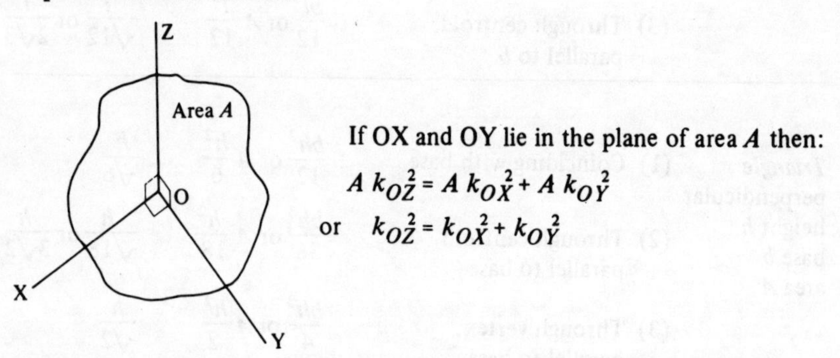

If OX and OY lie in the plane of area A then:

$$A\,k_{OZ}^2 = A\,k_{OX}^2 + A\,k_{OY}^2$$

or $$k_{OZ}^2 = k_{OX}^2 + k_{OY}^2$$

Differential Equations

First order differential equations

1. If $\dfrac{dy}{dx} = f(x)$ then $y = \displaystyle\int f(x)\,dx$

2. If $\dfrac{dy}{dx} = f(y)$ then $\displaystyle\int dx = \int \frac{dy}{f(y)}$

3. If $\dfrac{dy}{dx} = f(x).f(y)$ then $\displaystyle\int \frac{dy}{f(y)} = \int f(x)\,dx$

4. If $\dfrac{dQ}{dt} = kQ$ then $Q = Ae^{kt}$ (A and k are constants)

5. If $P\dfrac{dy}{dx} = Q$, where P and Q are functions of both x and y of the same degree throughout (i.e. a homogeneous first order differential equation), then

(i) rearrange into the form $\dfrac{dy}{dx} = \dfrac{Q}{P}$

(ii) make the substitution $y = vx$ (where v is a function of x), from which,

$$\frac{dy}{dx} = v(1) + x\frac{dv}{dx} \text{ (by the product rule)}$$

(iii) substitute for both y and $\dfrac{dy}{dx}$ in the equation $\dfrac{dy}{dx} = \dfrac{Q}{P}$

(iv) simplify, by cancelling, and an equation results in which the equations are separable

(v) separate the variables and solve using method 3 above

(vi) substitute $v = \dfrac{y}{x}$ to solve in terms of the original variable

6. If $\dfrac{dy}{dx} + Py = Q$, where P and Q are functions of x only (i.e. a linear first order differential equation), then

(i) determine the integrating factor, $e^{\int P dx}$

(ii) substitute the integrating factor (I.F.) into the equation

$$y(\text{I.F.}) = \int (\text{I.F.})Q\, dx$$

(iii) determine the integral $\int (\text{I.F.})Q\, dx$

Second order differential equations

7. If $a \frac{d^2y}{dx^2} + b \frac{dy}{dx} + cy = 0$ (where a, b and c are constants), then

(i) rewrite the differential equation as $(aD^2 + bD + c)y = 0$
(ii) substitute m for D and solve the auxiliary equation $am^2+bm+c = 0$
(iii) if the roots of the auxiliary equation are:
  (a) **real and different**, say $m = \alpha$ and $m = \beta$, then the general solution is

$$y = Ae^{\alpha x} + Be^{\beta x}$$

  (b) **real and equal**, say $m = \alpha$ twice, then the general solution is

$$y = (Ax+B)e^{\alpha x}$$

  (c) **complex**, say $m = \alpha \pm j\beta$, then the general solution is

$$y = e^{\alpha x} (A \cos \beta x + B \sin \beta x)$$

(iv) given boundary conditions, constants A and B can be determined and the particular solution obtained

8. If $a \frac{d^2y}{dx^2} + b \frac{dy}{dx} + cy = f(x)$ (where a, b and c are constants), then

(i) rewrite the differential equation as $(aD^2 + bD + c)y = f(x)$
(ii) substitute m for D and solve the auxiliary equation $am^2+bm+c = 0$
(iii) obtain the complementary function (C.F.), u, as per 7(iii) above
(iv) to find the particular integral, v, first assume a particular integral which is suggested by $f(x)$, but which contains undetermined coefficients. The table opposite gives some suggested substitutions.
(v) substitute the suggested particular integral into the original differential equation and equate relevant coefficients to find the constants introduced
(vi) the general solution is given by

$$y = u + v$$

(vii) given boundary conditions, arbitrary constants in the C.F. can be determined and the particular solution obtained.

Form of particular integral for different functions

Type	*Straightforward cases* Try as particular integral:	*'Snag' cases* Try as particular integral:
(a) $f(x)$ = a constant	$v = k$	$v = kx$ (used when C.F. contains a constant)
(b) $f(x)$ = a polynomial (i.e., $f(x) = L+Mx+Nx^2 + \ldots$) where any of the coefficients may be zero	$v = a+bx+cx^2 + \ldots$	
(c) $f(x)$ = an exponential function (i.e., $f(x) = Ae^{\alpha x}$)	$v = ke^{\alpha x}$	$v = kxe^{\alpha x}$ (used when $e^{\alpha x}$ appears in the C.F.) $v = kx^2e^{\alpha x}$ (used when $e^{\alpha x}$ **and** $xe^{\alpha x}$ both appear in the C.F.), and so on
(d) $f(x)$ = a sine or cosine function (i.e., $f(x) = a \sin px + b \cos px$, where a and b may be zero)	$v = A \sin px + B \cos px$	$v = x(A \sin px + B \cos px)$ (used when $\sin px$ and/or $\cos px$ appears in the C.F.)
(e) $f(x)$ = a sum e.g. (i) $f(x) = 2x^2+5 \cos 3x$ (ii) $f(x) = x+1-e^{-x}$	 (i) $v = ax^2+bx+c + d \cos 3x + e \sin 3x$ (ii) $v = ax+b+c\, e^{-x}$	
(f) $f(x)$ = a product e.g. $f(x) = 3e^{2x} \sin 4x$	 $v = e^{2x}(A \cos 4x + B \sin 4x)$	

Laplace transforms

Standard Laplace transforms

Function $f(t)$	Laplace transforms $\mathscr{L}\{f(t)\} = \int_0^\infty e^{-st} f(t)\,dt$
(1) 1	$\frac{1}{s}$
(2) k	$\frac{k}{s}$
(3) e^{at}	$\frac{1}{s-a}$
(4) $\sin at$	$\frac{a}{s^2+a^2}$
(5) $\cos at$	$\frac{s}{s^2+a^2}$
(6) t	$\frac{1}{s^2}$
(7) t^2	$\frac{2!}{s^3}$
(8) t^n (n = positive integer)	$\frac{n!}{s^{n+1}}$
(9) $\cosh at$	$\frac{s}{s^2-a^2}$
(10) $\sinh at$	$\frac{a}{s^2-a^2}$
(11) $e^{-at}t^n$	$\frac{n!}{(s+a)^{n+1}}$
(12) $e^{-at}\sin\omega t$	$\frac{\omega}{(s+a)^2+\omega^2}$
(13) $e^{-at}\cos\omega t$	$\frac{s+a}{(s+a)^2+\omega^2}$
(14) $e^{-at}\cosh\omega t$	$\frac{s+a}{(s+a)^2-\omega^2}$
(15) $e^{-at}\sinh\omega t$	$\frac{\omega}{(s+a)^2-\omega^2}$

The Laplace transforms of derivatives.

First derivative

$$\mathscr{L}\left\{\frac{dy}{dx}\right\} = s\,\mathscr{L}\{y\} - y(0),$$

when $y(0)$ is the value of y at $x = 0$

Second derivative

$$\mathscr{L}\left\{\frac{d^2y}{dx^2}\right\} = s^2\,\mathscr{L}\{y\} - s\,y(0) - y'(0)$$

where $y'(0)$ is the value of $\frac{dy}{dx}$ at $x = 0$.

Higher derivatives

$$\mathscr{L}\left\{\frac{d^ny}{dx^n}\right\} = s^n\,\mathscr{L}\{y\} - s^{n-1}y(0) - s^{n-2}y'(0) \ldots\ldots - y^{n-1}(0)$$

Fourier series

If $f(x)$ is a periodic function of period 2π then its Fourier series is given by:

$$f(x) = a_0 + \sum_{n=1}^{\infty}(a_n \cos nx + b_n \sin nx)$$

where, for the range $-\pi$ to $+\pi$

$$a_o = \frac{1}{2\pi}\int_{-\pi}^{\pi} f(x)dx$$

$$a_n = \frac{1}{\pi}\int_{-\pi}^{\pi} f(x)\cos nx\,dx \,(n = 1, 2, 3, \ldots)$$

$$b_n = \frac{1}{\pi}\int_{-\pi}^{\pi} f(x)\sin nx\,dx \,(n = 1, 2, 3, \ldots)$$

If $f(x)$ is a periodic function of period l then its Fourier series is given by:

$$f(x) = a_o + \sum_{n=1}^{\infty} \left\{ a_n \cos\left(\frac{2\pi nx}{l}\right) + b_n \sin\left(\frac{2\pi nx}{l}\right) \right\}$$

where, for the range $-\frac{l}{2}$ to $+\frac{l}{2}$

$$a_o = \frac{1}{l} \int_{-l/2}^{l/2} f(x)\, dx$$

$$a_n = \frac{2}{l} \int_{-l/2}^{l/2} f(x) \cos\left(\frac{2\pi nx}{l}\right) dx \qquad (n = 1, 2, 3, \ldots)$$

$$b_n = \frac{2}{l} \int_{-l/2}^{l/2} f(x) \sin\left(\frac{2\pi nx}{l}\right) dx \qquad (n = 1, 2, 3, \ldots)$$

Statistics

Grouped data

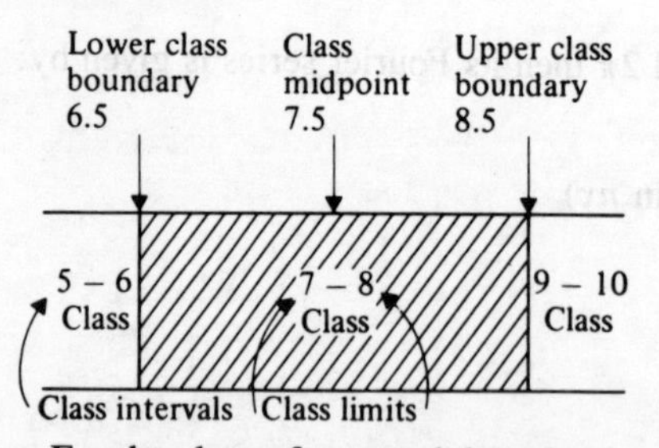

For the class of grouped data shown shaded, the class interval is from 7–8, the class limits are 7 and 8, the class midpoint is 7.5, the lower class boundary is 6.5 and the upper class boundary is 8.5.

Mean, median, mode and standard deviation

If x = variate and f = frequency then:

$$\text{mean, } \bar{x} = \frac{\Sigma f x}{\Sigma f}$$

The median is the middle term of a ranked set of data. The mode is the most commonly occurring value in a set of data.

$$\text{standard deviation, } \sigma = \sqrt{\left[\frac{\Sigma\{f(x - \bar{x})^2\}}{\Sigma f}\right]} \text{ for a population}$$

If $x, x_2, \ldots x_n$ is a random sample from a distribution having variance σ^2 then:

$$s^2 = \left(\frac{1}{n-1}\right) \sum_{i=1}^{n} (x_i - \bar{x})^2,$$

where $\bar{x}$ is the sample mean and s is the adjusted standard deviation of the sample.

Binomial probability distribution

If n = number in sample, p = probability of the occurrence of an event and $q = 1 - p$, then the probability of 0, 1, 2, 3, . . . occurrences is given by:

$$q^n,\ n q^{n-1} p,\ \frac{n(n-1)}{2!} q^{n-2} p^2,\ \frac{n(n-1)(n-2)}{3!} q^{n-3} p^3, \ldots$$

(i.e. successive terms of the $(q + p)^n$ expansion.)

Mean, $\mu = np$; Standard deviation, $\sigma = \sqrt{(n\,p\,q)}$

Poisson distribution

If λ is the expectation of the occurrence of an event then the probability of 0, 1, 2, 3, . . . occurrences is given by:

$$e^{-\lambda},\ \lambda e^{-\lambda},\ \lambda^2 \frac{e^{-\lambda}}{2!},\ \lambda^3 \frac{e^{-\lambda}}{3!}, \ldots$$

Mean, $\mu = \lambda = np$; Standard deviation, $\sigma = \sqrt{\lambda}$

Product-moment formula for the linear correlation coefficient

$$\text{Coefficient of correlation, } r = \frac{\Sigma xy}{\sqrt{[(\Sigma x^2)(\Sigma y^2)]}}$$

where $x = X - \bar{X}$ and $y = Y - \bar{Y}$ and $(X_1, Y_1), (X_2, Y_2), \ldots$ denote a random sample from a bivariate normal distribution and $\bar{X}$ and $\bar{Y}$ are the means of the X and Y values respectively.

Normal probability distribution.

Partial areas under the standardised normal curve.

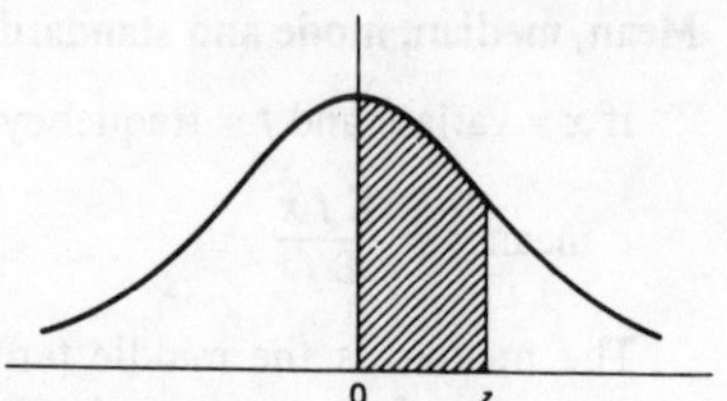

$z = \frac{x - \bar{x}}{\sigma}$	0	1	2	3	4	5	6	7	8	9
0.0	0.0000	0.0040	0.0080	0.0120	0.0159	0.0199	0.0239	0.0279	0.0319	0.0359
0.1	0.0398	0.0438	0.0478	0.0517	0.0557	0.0596	0.0636	0.0678	0.0714	0.0753
0.2	0.0793	0.0832	0.0871	0.0910	0.0948	0.0987	0.1026	0.1064	0.1103	0.1141
0.3	0.1179	0.1217	0.1255	0.1293	0.1331	0.1388	0.1406	0.1443	0.1480	0.1517
0.4	0.1554	0.1891	0.1628	0.1664	0.1700	0.1736	0.1772	0.1808	0.1844	0.1879
0.5	0.1915	0.1950	0.1985	0.2019	0.2054	0.2086	0.2123	0.2157	0.2190	0.2224
0.6	0.2257	0.2291	0.2324	0.2357	0.2389	0.2422	0.2454	0.2486	0.2517	0.2549
0.7	0.2580	0.2611	0.2642	0.2673	0.2704	0.2734	0.2760	0.2794	0.2823	0.2852
0.8	0.2881	0.2910	0.2939	0.2967	0.2995	0.3023	0.3051	0.3078	0.3106	0.3133
0.9	0.3159	0.3186	0.3212	0.3238	0.3264	0.3289	0.3215	0.3340	0.3365	0.3389
1.0	0.3413	0.3438	0.3451	0.3485	0.3508	0.3531	0.3554	0.3577	0.3599	0.3621
1.1	0.3643	0.3665	0.3686	0.3708	0.3729	0.3749	0.3770	0.3790	0.3810	0.3830
1.2	0.3849	0.3869	0.3888	0.3907	0.3925	0.3944	0.3962	0.3980	0.3997	0.4015
1.3	0.4032	0.4049	0.4066	0.4082	0.4099	0.4115	0.4131	0.4147	0.4162	0.4177
1.4	0.4192	0.4207	0.4222	0.4236	0.4251	0.4265	0.4279	0.4292	0.4306	0.4319
1.5	0.4332	0.4345	0.4357	0.4370	0.4382	0.4394	0.4406	0.4418	0.4430	0.4441
1.6	0.4452	0.4463	0.4474	0.4484	0.4495	0.4505	0.4515	0.4525	0.4535	0.4545
1.7	0.4554	0.4564	0.4573	0.4582	0.4591	0.4599	0.4608	0.4616	0.4625	0.4633
1.8	0.4641	0.4649	0.4656	0.4664	0.4671	0.4678	0.4686	0.4693	0.4699	0.4706
1.9	0.4713	0.4719	0.4726	0.4732	0.4738	0.4744	0.4750	0.4756	0.4762	0.4767
2.0	0.4772	0.4778	0.4783	0.4785	0.4793	0.4798	0.4803	0.4808	0.4812	0.4817
2.1	0.4821	0.4826	0.4830	0.4834	0.4838	0.4842	0.4846	0.4850	0.4854	0.4857
2.2	0.4861	0.4864	0.4868	0.4871	0.4875	0.4878	0.4881	0.4884	0.4882	0.4890
2.3	0.4893	0.4896	0.4898	0.4901	0.4904	0.4906	0.4909	0.4911	0.4913	0.4916
2.4	0.4918	0.4920	0.4922	0.4925	0.4927	0.4929	0.4931	0.4932	0.4934	0.4936
2.5	0.4938	0.4940	0.4941	0.4943	0.4945	0.4946	0.4948	0.4949	0.4951	0.4952
2.6	0.4953	0.4955	0.4956	0.4957	0.4959	0.4960	0.4961	0.4962	0.4963	0.4964
2.7	0.4965	0.4966	0.4967	0.4968	0.4969	0.4970	0.4971	0.4972	0.4973	0.4974
2.8	0.4974	0.4975	0.4976	0.4977	0.4977	0.4978	0.4979	0.4980	0.4980	0.4981
2.9	0.4981	0.4982	0.4982	0.4983	0.4984	0.4984	0.4985	0.4985	0.4986	0.4986
3.0	0.4987	0.4987	0.4987	0.4988	0.4988	0.4989	0.4989	0.4989	0.4990	0.4990
3.1	0.4990	0.4991	0.4991	0.4991	0.4992	0.4992	0.4992	0.4992	0.4993	0.4993
3.2	0.4993	0.4993	0.4994	0.4994	0.4994	0.4994	0.4994	0.4995	0.4995	0.4995
3.3	0.4995	0.4995	0.4995	0.4996	0.4996	0.4996	0.4996	0.4996	0.4996	0.4997
3.4	0.4997	0.4997	0.4997	0.4997	0.4997	0.4997	0.4997	0.4997	0.4997	0.4998
3.5	0.4998	0.4998	0.4998	0.4998	0.4998	0.4998	0.4998	0.4998	0.4998	0.4998
3.6	0.4998	0.4998	0.4999	0.4999	0.4999	0.4999	0.4999	0.4999	0.4999	0.4999
3.7	0.4999	0.4999	0.4999	0.4999	0.4999	0.4999	0.4999	0.4999	0.4999	0.4999
3.8	0.4999	0.4999	0.4999	0.4999	0.4999	0.4999	0.4999	0.4999	0.4999	0.4999
3.9	0.5000	0.5000	0.5000	0.5000	0.5000	0.5000	0.5000	0.5000	0.5000	0.5000

Student's *t* distribution

Percentile values (t_p) for Student's *t* distribution with v degrees of freedom (shaded area = p)

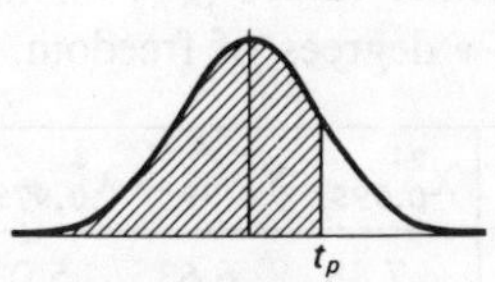

v	$t_{0.995}$	$t_{0.99}$	$t_{0.975}$	$t_{0.95}$	$t_{0.90}$	$t_{0.80}$	$t_{0.75}$	$t_{0.70}$	$t_{0.60}$	$t_{0.55}$
1	63.66	31.82	12.71	6.31	3.08	1.376	1.000	0.727	0.325	0.158
2	9.92	6.96	4.30	2.92	1.89	1.061	0.816	0.617	0.289	0.142
3	5.84	4.54	3.18	2.35	1.64	0.978	0.765	0.584	0.277	0.137
4	4.60	3.75	2.78	2.13	1.53	0.941	0.741	0.569	0.271	0.134
5	4.03	3.36	2.57	2.02	1.48	0.920	0.727	0.559	0.267	0.132
6	3.71	3.14	2.45	1.94	1.44	0.906	0.718	0.553	0.265	0.131
7	3.50	3.00	2.36	1.90	1.42	0.896	0.711	0.549	0.263	0.130
8	3.36	2.90	2.31	1.86	1.40	0.889	0.706	0.546	0.262	0.130
9	3.25	2.82	2.26	1.83	1.38	0.883	0.703	0.543	0.261	0.129
10	3.17	2.76	2.23	1.81	1.37	0.879	0.700	0.542	0.260	0.129
11	3.11	2.72	2.20	1.80	1.36	0.876	0.697	0.540	0.260	0.129
12	3.06	2.68	2.18	1.78	1.36	0.873	0.695	0.539	0.259	0.128
13	3.01	2.65	2.16	1.77	1.35	0.870	0.694	0.538	0.259	0.128
14	2.98	2.62	2.14	1.76	1.34	0.868	0.692	0.537	0.258	0.128
15	2.95	2.60	2.13	1.75	1.34	0.866	0.691	0.536	0.258	0.128
16	2.92	2.58	2.12	1.75	1.34	0.865	0.690	0.535	0.258	0.128
17	2.90	2.57	2.11	1.74	1.33	0.863	0.689	0.534	0.257	0.128
18	2.88	2.55	2.10	1.73	1.33	0.862	0.688	0.534	0.257	0.127
19	2.86	2.54	2.09	1.73	1.33	0.861	0.688	0.533	0.257	0.127
20	2.84	2.53	2.09	1.72	1.32	0.860	0.687	0.533	0.257	0.127
21	2.83	2.52	2.08	1.72	1.32	0.859	0.686	0.532	0.257	0.127
22	2.82	2.51	2.07	1.72	1.32	0.858	0.686	0.532	0.256	0.127
23	2.81	2.50	2.07	1.71	1.32	0.858	0.685	0.532	0.256	0.127
24	2.80	2.49	2.06	1.71	1.32	0.857	0.685	0.531	0.256	0.127
25	2.79	2.48	2.06	1.71	1.32	0.856	0.684	0.531	0.256	0.127
26	2.78	2.48	2.06	1.71	1.32	0.856	0.684	0.531	0.256	0.127
27	2.77	2.47	2.05	1.70	1.31	0.855	0.684	0.531	0.256	0.127
28	2.76	2.47	2.05	1.70	1.31	0.855	0.683	0.530	0.256	0.127
29	2.76	2.46	2.04	1.70	1.31	0.854	0.683	0.530	0.256	0.127
30	2.75	2.46	2.04	1.70	1.31	0.854	0.683	0.530	0.256	0.127
40	2.70	2.42	2.02	1.68	1.30	0.851	0.681	0.529	0.255	0.126
60	2.66	2.39	2.00	1.67	1.30	0.848	0.679	0.527	0.254	0.126
120	2.62	2.36	1.98	1.66	1.29	0.845	0.677	0.526	0.254	0.126
∞	2.58	2.33	1.96	1.645	1.28	0.842	0.674	0.524	0.253	0.126

Chi-square distribution

Percentile values (χ_p^2) for the Chi-square distribution with ν degrees of freedom.

ν	$\chi^2_{0.995}$	$\chi^2_{0.99}$	$\chi^2_{0.975}$	$\chi^2_{0.95}$	$\chi^2_{0.90}$	$\chi^2_{0.75}$	$\chi^2_{0.50}$	$\chi^2_{0.25}$
1	7.88	6.63	5.02	3.84	2.71	1.32	0.455	0.102
2	10.6	9.21	7.38	5.99	4.61	2.77	1.39	0.575
3	12.8	11.3	9.35	7.81	6.25	4.11	2.37	1.21
4	14.9	13.3	11.1	9.49	7.78	5.39	3.36	1.92
5	16.7	15.1	12.8	11.1	9.24	6.63	4.35	2.67
6	18.5	16.8	14.4	12.6	10.6	7.84	5.35	3.45
7	20.3	18.5	16.0	14.1	12.0	9.04	6.35	4.25
8	22.0	20.1	17.5	15.5	13.4	10.2	7.34	5.07
9	23.6	21.7	19.0	16.9	14.7	11.4	8.34	5.90
10	25.2	23.2	20.5	18.3	16.0	12.5	9.34	6.74
11	26.8	24.7	21.9	19.7	17.3	13.7	10.3	7.58
12	28.3	26.2	23.3	21.0	18.5	14.8	11.3	8.44
13	29.8	27.7	24.7	22.4	19.8	16.0	12.3	9.30
14	31.3	29.1	26.1	23.7	21.1	17.1	13.3	10.2
15	32.8	30.6	27.5	25.0	22.3	18.2	14.3	11.0
16	34.3	32.0	28.8	26.3	23.5	19.4	15.3	11.9
17	35.7	33.4	30.2	27.6	24.8	20.5	16.3	12.8
18	37.2	34.8	31.5	28.9	26.0	21.6	17.3	13.7
19	38.6	36.2	32.9	30.1	27.2	22.7	18.3	14.6
20	40.0	37.6	34.4	31.4	28.4	23.8	19.3	15.5
21	41.4	38.9	35.5	32.7	29.6	24.9	20.3	16.3
22	42.8	40.3	36.8	33.9	30.8	26.0	21.3	17.2
23	44.2	41.6	38.1	35.2	32.0	27.1	22.3	18.1
24	45.6	43.0	39.4	36.4	33.2	28.2	23.3	19.0
25	46.9	44.3	40.6	37.7	34.4	29.3	24.3	19.9
26	48.3	45.9	41.9	38.9	35.6	30.4	25.3	20.8
27	49.6	47.0	43.2	40.1	36.7	31.5	26.3	21.7
28	51.0	48.3	44.5	41.3	37.9	32.6	27.3	22.7
29	52.3	49.6	45.7	42.6	39.1	33.7	28.3	23.6
30	53.7	50.9	47.7	43.8	40.3	34.8	29.3	24.5
40	66.8	63.7	59.3	55.8	51.8	45.6	39.3	33.7
50	79.5	76.2	71.4	67.5	63.2	56.3	49.3	42.9
60	92.0	88.4	83.3	79.1	74.4	67.0	59.3	52.3
70	104.2	100.4	95.0	90.5	85.5	77.6	69.3	61.7
80	116.3	112.3	106.6	101.9	96.6	88.1	79.3	71.1
90	128.3	124.1	118.1	113.1	107.6	98.6	89.3	80.6
100	140.2	135.8	129.6	124.3	118.5	109.1	99.3	90.1

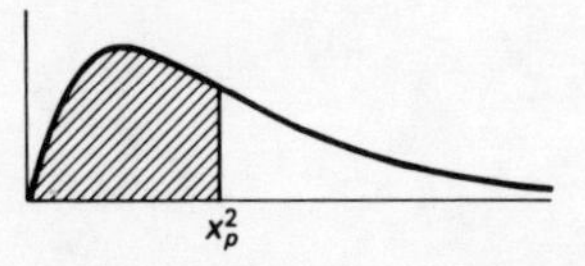

$\chi^2_{0.10}$	$\chi^2_{0.05}$	$\chi^2_{0.025}$	$\chi^2_{0.01}$	$\chi^2_{0.005}$
0.0158	0.0039	0.0010	0.0002	0.0000
0.211	0.103	0.0506	0.0201	0.0100
0.584	0.352	0.216	0.115	0.072
1.06	0.711	0.484	0.297	0.207
1.61	1.15	0.831	0.554	0.412
2.20	1.64	1.24	0.872	0.676
2.83	2.17	1.69	1.24	0.989
3.49	2.73	2.18	1.65	1.34
4.17	3.33	2.70	2.09	1.73
4.87	3.94	3.25	2.56	2.16
5.58	4.57	3.82	3.05	2.60
6.30	5.23	4.40	3.57	3.07
7.04	5.89	5.01	4.11	3.57
7.79	6.57	5.63	4.66	4.07
8.55	7.26	6.26	5.23	4.60
9.31	7.96	6.91	5.81	5.14
10.1	8.67	7.56	6.41	5.70
10.9	9.39	8.23	7.01	6.26
11.7	10.1	8.91	7.63	6.84
12.4	10.9	9.59	8.26	7.43
13.2	11.6	10.3	8.90	8.03
14.0	12.3	11.0	9.54	8.64
14.8	13.1	11.7	10.2	9.26
15.7	13.8	12.4	10.9	9.89
16.5	14.6	13.1	11.5	10.5
17.3	15.4	13.8	12.2	11.2
18.1	16.2	14.6	12.9	11.8
18.9	16.9	15.3	13.6	12.5
19.8	17.7	16.0	14.3	13.1
20.6	18.5	16.8	15.0	13.8
29.1	26.5	24.4	22.2	20.7
37.7	34.8	32.4	29.7	28.0
46.5	43.2	40.5	37.5	35.5
55.3	51.7	48.8	45.4	43.3
64.3	60.4	57.2	53.5	51.2
73.3	69.1	65.6	61.8	59.2
82.4	77.9	74.2	70.1	67.3